WHERE WE LIVE OUR SOLAR SYSTEM

THE THIRD PLANET FROM THE SUN!

BY JONI MORTON

Table of Contents

1. Our Solar System
2. Sun – Ancient Romans called it Sol, the Greeks - Helios
3. Mercury – The Messenger
4. Venus – Goddess of Love
5. Earth – 1st Goddess on earth (Terra Mater) Greeks-(Gaeato)
6. Mars –God of war
7. Ceres – Goddess of corn and harvest
8. Astroid Belt – Kuiper Belt – Ort Cloud
9. Beyond the Ort Cloud
10. Jupiter – Sky God - The Greeks call is Zeus Saturn – God   of sowing seeds and harvest
11. Saturn – God of sowing seed and Harvest
12. Comets – Greek name Kometes – (long hair)
13. Uranus – God of Heavens
14. Neptune – God of Sea
15. Pluto – God of the Underworld
16. MakeMake – Polynesian Creator God
17. Eris –Goddess of Discard and Strife

Our Solar System

The Solar System is made up of all the planets that orbit our Sun. In addition to planets, the Solar System also consists of moons, comets, asteroids, minor planets, and dust and gas.

Image of Our Solar System

Our Sun

Everything in the Solar System orbits or revolves around the Sun. The Sun contains around 98% of all the material in the Solar System. The larger an object is, the more gravity it has. Because the Sun is so large, its powerful gravity attracts all the other objects in the Solar System towards it. At the same time, these objects, which are moving very rapidly, try to fly away from the Sun, outward into the emptiness of outer space. The result of the planets trying to fly away, at the same time that the Sun is trying to pull them inward is that they become trapped half-way in between. Balanced between flying towards the Sun, and escaping into space, they spend eternity orbiting around their parent star.

(Photo of our sun)

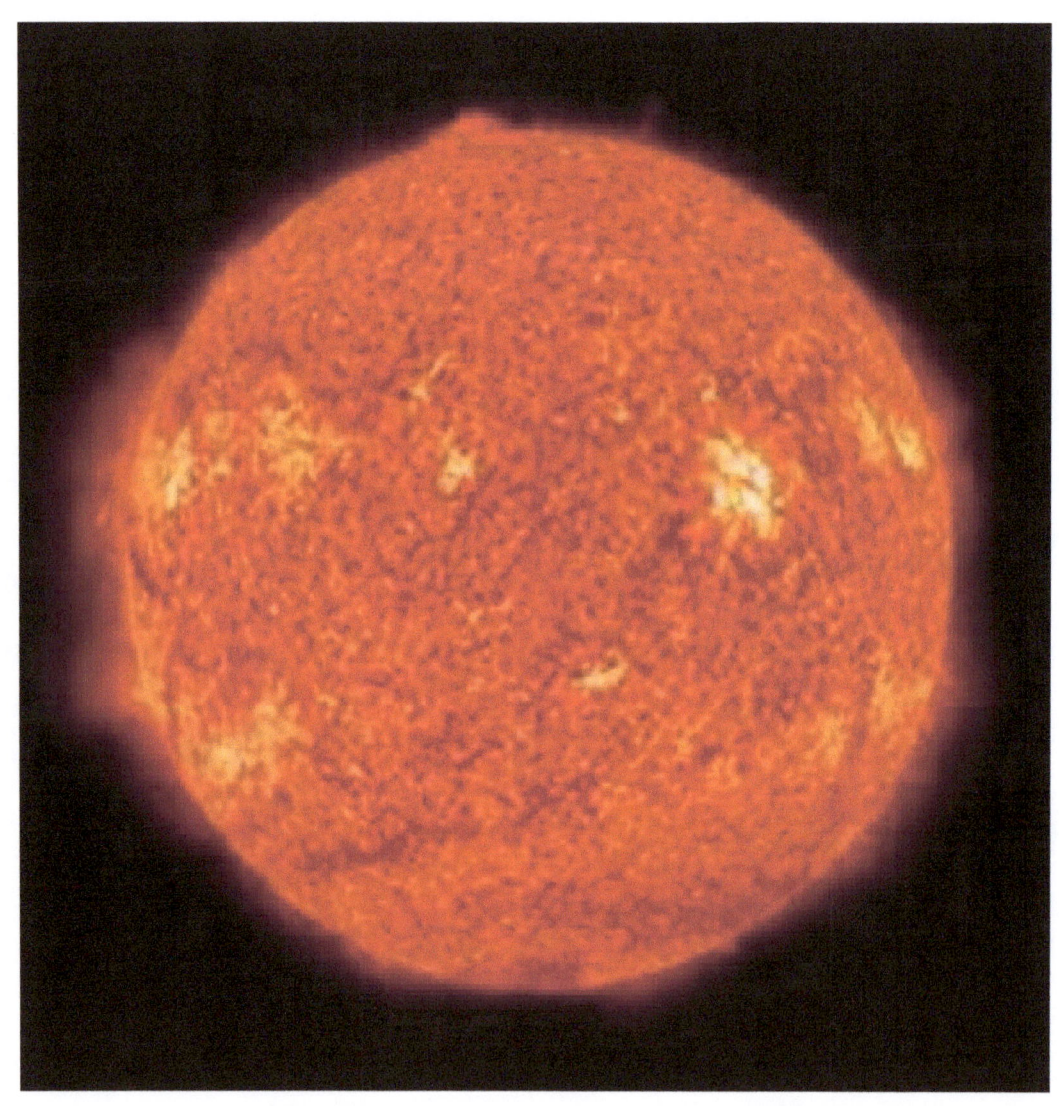

Mercury

Mercury is the planet closest to the Sun. It is not, however, very close, since it is 36 million miles, or 58 million kilometers away from the sun.

The distances of planets from each other and from the Sun are often measured in Astronomical Units, AU. One AU is the average distance between the Earth and the Sun, 93 million miles, or 150 million kilometers. Using this system of measurement, Mercury is 0.39 AU from the Sun. Like all the other planets Mercury orbits round the Sun, but its orbit of the Sun lasts for only 88 days. The earths orbit lasts for 365 days and Pluto's orbit takes 249 YEARS. Because Mercury goes round the Sun so quickly, the planet was called after the messenger of the Roman Gods. The messenger Mercury, or Hermes as the Greeks knew him, is usually shown as having wings on his helmet or on his sandals. When Mercury orbits the Sun, it travels 36 million miles, or 58 million kilometers in the 88 days of the orbit. It moves at a speed of 48 kilometers a second, or 107,372 miles an hour. Unlike the Earth and most other planet Mercury only turns very slowly on its axis, taking 59 days to complete the turn from day to night. Mercury's sunny side has a temperature rising to 400° Celsius or 750° Fahrenheit. Compare this to a warm summer's day in London, when the temperature might be 80° Fahrenheit or 26° Celsius. Mercury's dark side however, is very cold indeed, with the temperature going down to -200° Celsius or -328° Fahrenheit. Mercury has no atmosphere around it to protect it from the Sun or to retain any heat when it rotates on its axis. Mercury is quite a small planet. Its diameter, the distance right round its middle, is only 3100 miles or 4990 kilometers. The diameter of the Earth is 7926 miles or 12,760 kilometers. Mercury's distance from the Earth is 57 million miles, or 92 million kilometers. Using Astronomical Units Mercury is 0.61 AU from the Earth. Mercury has no moons. Moons are satellites that travel with a planet as it orbits the sun. The earth has one moon, Mars has two very small ones, Jupiter, the giant of the planets, has 16. The surface of Mercury is covered with craters and completely dry.

There is no possibility of life on Mercury. The first photographs of the surface of Mercury were taken by the USA Space Agency, NASA.

The Mariner 10 spacecraft passed close to the planet in 1974 and 1975 and took very clear photographs. NASA's latest mission to Mercury is called Messenger. The Messenger spacecraft entered Mercury's orbit in March 2011 and is sending back new pictures of the planet. Messenger is now moving with Mercury round the Sun. From Messenger we know that Mercury has a large number of very deep and irregular pits. Some of these pits are several miles deep. Mercury is one of five planets that can be seen without using a telescope, Mercury, Venus, Mars, Jupiter and Saturn. When you look at the sky at night, the planets do not twinkle in the way that stars do. Mercury is not very easy to see, but it can be seen low in the west just after sunset or in the east just before dawn. About once every ten or fifteen years Mercury can be seen crossing the Sun. At this point its orbit has come between the Sun and the Earth.

This event is known as a transit. When watching any event near the Sun a proper filter must be used to protect the sight. With this filter Mercury can be seen as a tiny black dot slowly passing across the Sun. You must never try to look directly at the Sun without a filter.

(Photo of Mercury)

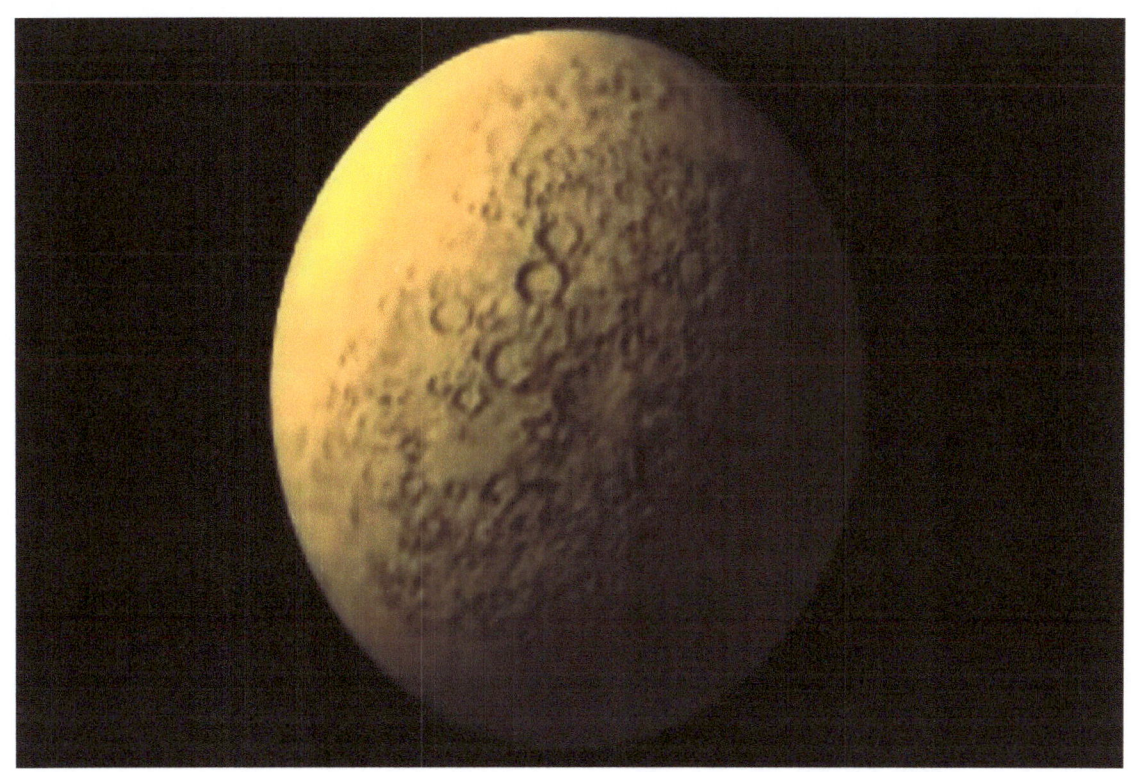

Venus

Venus is the brightest planet in the Solar System and can be seen even in daylight if you know where to look. When Venus is west of the Sun, she rises before the Sun in the morning and is known as the Morning Star. When she is east of the Sun, she shines in the evening just after sunset and is known as the Evening Star. The word planet is from the ancient Greek word (planets), which means a wanderer, because they move through the stars, which seem to be fixed in the sky. This movement is because the planets are all orbiting around the sun. Venus orbits round the sun in 225 days. The Earth takes 365 days to complete an orbit of the sun. So a year on Venus only lasts for 262 days! Venus has phases like the moon because the orbit of Venus is between the Earth and the Sun. When Venus shows only a crescent, like the crescent moon, she is at her brightest because she is then very close to the Earth. You can only see the crescent with the help of a telescope.

Venus is the 2nd planet from the sun. The closest planet to the sun is Mercury. Then comes Venus and then Earth, followed by Mars,

Saturn, Jupiter, Uranus, Neptune and Pluto. There are other orbiting bodies beyond Pluto but astronomers scientists who study the Solar System do not agree about whether these are planets. The planets in the Solar system are given the names of Roman Gods or their attendants. Venus is called after the Roman Goddess of love and beauty. Venus is the planet which is closest to the Earth and is a little smaller than the Earth. The circumference of the Earth the distance right round the middle of the Earth at the equator is 12,760 kilometers, or 7926 miles. The circumference of Venus is 12,103 kilometers, or 7520 miles. Venus looks bigger than the Earth because of the very deep layer of gases that surround the planet. Venus, like the other planet between the Earth and the Sun, Mercury, has no moons. Venus is 67 million miles, or 108 million kilometers, from the Sun. Venus is covered by clouds of water vapor and sulfuric acid and the surface cannot be seen with an ordinary astronomy telescope. Venus is the hottest planet in the Solar System, even hotter than Mercury, which is closer to the Sun. The temperature on the surface of Venus is about 860° Fahrenheit or 460° Celsius. Compare this to a warm summer's day in Texas, when the temperature might be 80° Fahrenheit or 26° Celsius.

The atmosphere on Venus is composed of carbon dioxide. The surface is heated by radiation from the sun, but the heat cannot escape through the clouds and layer of carbon dioxide. This is a greenhouse effect. In 1975 the Soviet Union sent two spacecraft landers to Venus, Venera 9 and Venera 10. These landers were the very first to reach the planet's surface and they sent back images to the Earth. Venus has continents, mountains and craters.

There is no life as we know it at all on Venus and life could never be supported there because of the extreme heat and the atmosphere. The mountains and craters on Venus have all been given female names such as the crater called Billie Holiday after a female American jazz singer. There is only one male name the Maxwell Mountains which are called after the Scottish scientist James Clerk Maxwell. The largest continent on Venus has been named Aphrodite. Aphrodite was the Greek Goddess of love and beauty and is really the same deity as Venus.

(Photo of Venus)

Earth

The Earth is unlike every other planet in the Solar System in a number of different ways. It is the only planet that has an atmosphere containing 21 percent oxygen. To date it is the only planet that has liquid water on its surface. It is the only planet in the solar system that has life. That we know of. The Earth is the only inner planet (Mercury, Venus, Earth and Mars) to have one large satellite, the Moon. Mars has two very tiny moons. Mercury and Venus have none. The Earth is fragile. Its surface is split into plates (tectonic plates) which float on a rocky mantle – the layer between the surface of the earth, its crust, and its hot liquid core. The inside of the Earth is active and earthquakes, volcanoes and mountain building takes place along the boundaries of the tectonic plates. As a result of the Earth's geological activity (the volcanoes and earthquakes) the surface of the Earth has far fewer craters than the surface of planets such as Mars, Venus and Mercury or the surface of the Moon.

The craters have sunk down or been worn away by wind and rain over millions of years. When viewed from outer space much of the Earth's surface cannot be seen because of clouds of water vapor. The water vapor makes the Earth, when seen from outside, into a brilliant shining orb. The Earth is the third planet from the Sun and comes between the planet Venus and the planet Mars. The Earth takes 365¼ days to complete its orbit round the Sun. The Earth's year is therefore 365 days long but the ¼ days are added up and every fourth year has one extra day, on the 29th of February. This fourth year is called a Leap Year (366 days) and is always a year which can be divided exactly by 4 – 2000, 2004, 2008, 2012, 2016. The planets closer to the Sun, Mercury and Venus, have shorter years than the Earth. The planets further away from the Sun have longer years; Pluto takes 249 of our years to make one orbit of the Sun. As the Earth orbits round the Sun it turns on its axis, rotating right round in 24 hours. The side of the Earth that faces the Sun has daytime and the side of the Earth that is turned away from the Sun has night-time. When it is daytime in Britain, it is night-time on the opposite side of the Earth in New Zealand. As the Earth orbits round the Sun it tilts very slightly and so gives us the seasons. When the Earth has tilted so that the northern half of the Earth is a little away from the Sun, the northern hemisphere (meaning half of the Earth's sphere) has winter. At this time the southern hemisphere is tilted very slightly towards the Sun and the southern hemisphere has summer. Winter in Britain means summer in New Zealand. Closer to the Equator there is much less difference between summer and winter. The Earth is 93 million miles, or 150 million kilometers from the Sun.

The Earth's diameter, the distance round its middle at the Equator, is 7928 miles, or 12760 kilometers. The Earth is not an exact sphere; the diameter going round the North and South Poles is slightly less than the diameter round the Equator. The Polar diameter is 7891 miles, or 12700 kilometers. The Earth is larger than Mercury, Venus and Mars, the planets closest to it.

The Earth differs from all the other planets because it has such a wide diversity of life and intelligent beings. This has only been possible because of the Earth's atmosphere which has protected the Earth and allowed life to flourish.

(Photo of Earth)

Mars

It is easy to forget that Earth is not the only planet in the solar system. eight other planets whiz around the sun just like ours. Of those planets, none of them are closer or more engaging to the imagination than Mars. The Red Planet, as Mars is often called, is the fourth planet from the sun (Earth is the third). In a lot of ways, Mars looks a lot like our home, though instead of blue oceans and green land, Mars is home to an ever present red tint. This is due to a mineral called iron oxide that is very common on the planet's surface. However, when you look past the surface differences, these two planets are similar in a lot of ways. Here are just a few: Polar Ice Caps – Mars has both North and South polar ice caps, much like Earth. Also like Earth, both ice caps are made mostly of frozen water. With so much water frozen in the ice caps of Mars, some scientists think that life could have once existed there. Mars is not much farther away from the Sun than Earth. As a result, a typical year on Mars is 1 year and 320 days. Length of a Day – While a year on Mars might be almost twice as long as a year on Earth, the length of a day there is almost identical. A Martian day is 24 hours and 39 minutes long, less than an hour longer than a day on Earth. Mars has seasons like Earth too. These seasons are much longer than Earth seasons because Mars is so much farther from the sun. The average high during a Martian summer day is 23 degrees F (-5 degrees C) Mars and Earth are similar in so many ways that it's almost hard to believe we haven't found anything alive there. But, don't forget that there are many differences too. Without these differences, Mars wouldn't be such an interesting planet to study.

A lot of planets are bigger than Earth. For example, 318 Earths could fit inside of Jupiter. Mars is not quite so big. In fact, Mars is one of only two planets in the solar system to be significantly smaller than Earth. If you looked at the two planets side by side, Earth would be a basketball while Mars is a softball. But, just because Mars is smaller doesn't mean it is without landmarks. The surface of Mars is filled with exciting locations,

Olympus Mons on Mars is the largest mountain in the solar system at more than 25 kilometers high that is three times higher than Mount Everest. It is so big that astronomers could see it through telescopes almost 200 years ago! Valles Marineris is the largest canyon in the solar system, stretching 4,000 kilometers across the planet's surface. If you look at a picture of Mars taken from a telescope, you will see the giant gash that is Valles Marineris. The Borealis Basin makes up 40% of the planet's surface, taking up almost the entire northern hemisphere.

Mars is covered by craters from objects like asteroids and meteorites hitting the planet. Today, 43,000 such craters have been found and that only includes the large ones! Mars doesn't have a protective layer of atmosphere like Earth, so it cannot store heat from the sun. As a result, the temperature on Mars regularly drops to -125 degrees F (-82 degrees C) in the winter and only rises to 23 degrees F (-5 degrees C) in the summer.

The dust storms on Mars are larger than on any other planet in the solar system. Some dust storms on Mars can blanket almost the entire planet in just a few days Mars is an incredible planet. With mountains, craters and caverns like Earth and a rich history, we will be learning more about the Red Planet for centuries to come.

(Photo of Mars)

Ceres

Ceres was the first object considered to be an asteroid. Italian astronomer Giuseppe Piazzi discovered and named Ceres in early 1801. The first visit to Ceres is due in 2015 Nasa's Dawn spacecraft has been making its way to Ceres from the asteroid Vesta since September 2012. There is high interest in this mission since Ceres will be the first Dwarf Planet visited by a spacecraft and is one possible destination for human colonization given its abundance of ice, water, and minerals. Ceres has a mysterious white spot. This can be seen in both the old Hubble images and the more recent photos taken by the Dawn spacecraft on its approach. Every second Ceres loses 6kg of its mass in steam. Plumes of water vapor shooting up from Ceres' surface were observed by the Herschel Space Telescope this was the first definitive observation of water vapor in the asteroid belt. It's thought this is caused when portions of Ceres' icy surface warm. Ceres accounts for one third of the mass in the asteroid belt. Despite this it is still the smallest and least massive of the dwarf planets. For roughly the first 50 years after its discovery Ceres was frequently referred to as a planet. By the end of 1851, 14 other similar objects had been discovered and it did not take long before these instead became known as minor planets. Ceres was eventually reclassified as a Dwarf Planet alongside Pluto in 2006.

(Photo of Ceres)

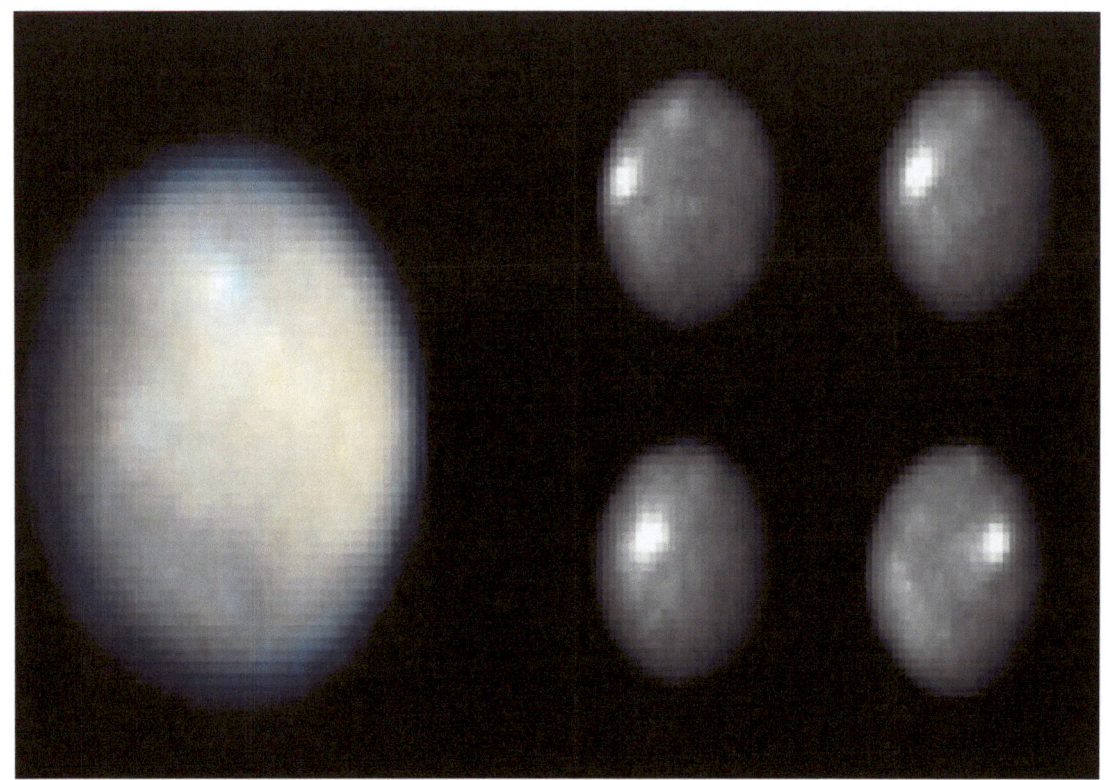

The Asteroid Belt, The Kuiper Belt, And The Oort Cloud

You have probably heard about the Asteroid Belt. This band of asteroids sits between the orbits of the planets Jupiter and Mars. It is made up of thousands of objects too small to be considered planets. Some of them no larger than a grain of dust, while others, like Eros can be more than 100 miles across. A few, like Ida, even have their own moons. Further out, beyond the orbit of the minor planet Pluto, sits another belt known as the Kuiper Belt.

Like the Asteroid Belt, the Kuiper Belt is also made up of thousands, possibly even millions of objects too small to be considered planets. A few of these objects, like Pluto, are large enough that their gravity has pulled them into a sphere shape.

These objects are made out of mostly frozen gas with small amounts of dust. They are often called dirty snowballs. However, you probably know them by their other name comets.

Every once in a while one of these comets will be thrown off of its orbit in the Kuiper Belt and hurled towards the inner Solar

System where it slowly melts in a fantastic show of tail and light. Beyond the Kuiper Belt sits a vast area known as the Oort Cloud. Here within this jumbled disorganized cloud live millions of additional comets. These comets do not orbit the Sun in a ring or belt. Instead, each one buzzes around in a completely random direction, and at extremely high velocities.

(Photo of Kiuper Belt)

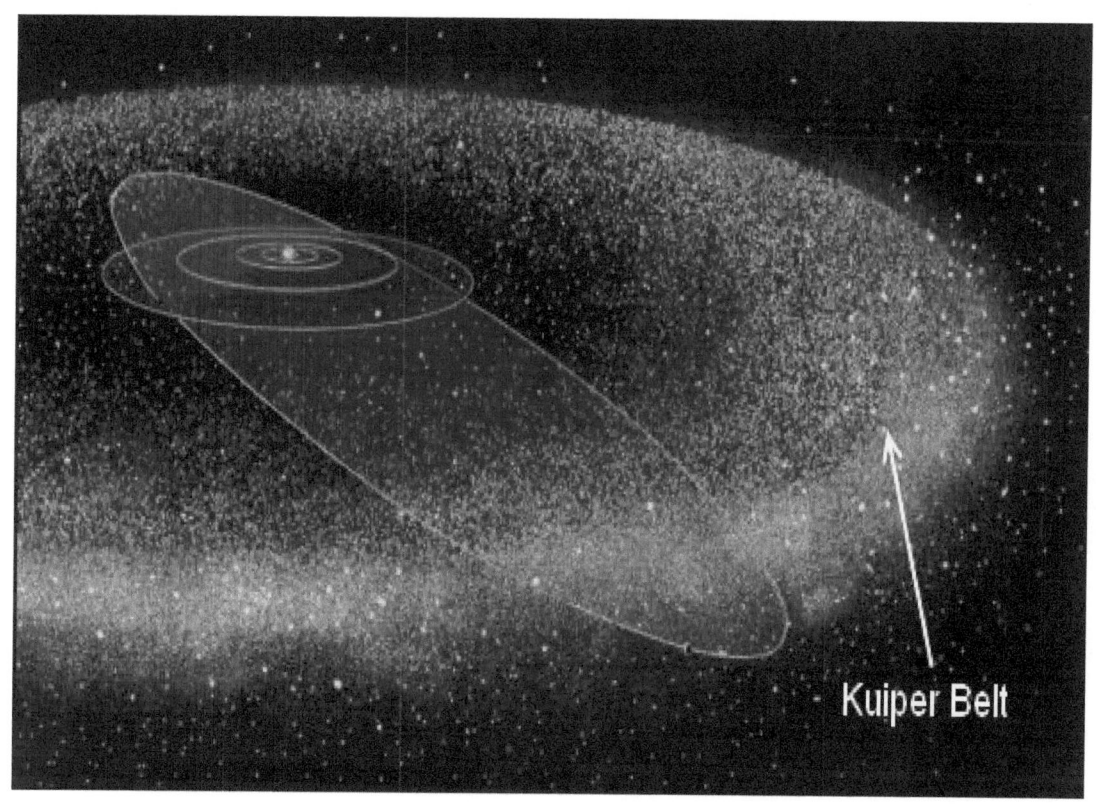
Kuiper Belt

(Photo of the Asteroid Belt)

Beyond The Oort Cloud

The Sun's solar winds continue pushing outward until they finally begin to mix into the interstellar medium, becoming lost with the winds from other stars. This creates a sort of bubble called the Heliosphere. Scientists define the boundaries of the Solar System as being the border of the Heliosphere, or at the place where the solar winds from the Sun mix with the winds from other stars. The Heliosphere extends out from the Sun to a distance of about 15 billion miles, which is more than 160 times further from the Sun than is the Earth.

Jupiter

Jupiter is the giant of the Solar System, with a mass more than 300 times the mass of the Earth and is called after the ancient Roman sky-god, Jupiter, known to the Greeks as Zeus. Jupiter has a diameter of 88,700 miles, or 142,750 kilometers. Jupiter is the fifth planet in order from the Sun and is about 483 million miles, or 777 million kilometers from the Sun. The Earth is much closer to the Sun than it is to Jupiter. Although Jupiter's orbit, and therefore its year, is so much longer than the Earth's, its day is much shorter. The Earth turns on its own axis, turning away from the Sun and so giving us day and night, once every 24 hours. Jupiter spins round much faster, turning on its axis once every 9.84 hours! This fast spinning gives rise to very strong weather patterns in the clouds which surround the planet and
so its appearance changes rapidly. Jupiter is the stormiest planet in the Solar System. There is a permanent, but ever-changing whirlpool of storms, known as Jupiter's Great Red Spot which can be seen using a telescope. The Red Spot was first seen by

Robert Hooke in 1664. Jupiter is the first of the gas giants, Jupiter, Saturn, Uranus and Neptune. The gas giants are entirely composed of dense layers of gas. Jupiter is made of hydrogen, helium, methane and ammonia. The cloudy sphere has bright belts on it which change their shape.

Jupiter can be seen without a telescope and so was known in the ancient world, but it was not until the invention of telescopes that astronomers were able to see Jupiter's moons. There are 64 moons in total, four of which are large enough to be easily observed with a small telescope. The first person to discover and observe Jupiter's moons was Galileo (1564-1642). Closest to Jupiter is Io, further away is Europa, and there are two large outer moons, Ganymede and Callisto. Io is so close to the planet that the pull of Jupiter's gravity is constantly disturbing Io's surface with volcanic eruptions.

Europa is coated with smooth ice, while Ganymede and Callisto both have much older ice, deeply pitted with craters.

There is still much to be learned about Jupiter. On 5th August 2011

NASA, launched the Juno an unmanned spacecraft on a mission to Jupiter to try to learn more. Juno was a Roman goddess and the wife of the god Jupiter. The spacecraft Juno will take five years to reach the planet Jupiter. It is the first spacecraft to be solar-powered.

(Photo of Jupiter)

Saturn

Saturn is the sixth planet in the Solar system and, when seen through a telescope, by far the most beautiful.

The bright globe of Saturn is surrounded by rings which may be composed of ice. Three of these rings are visible from the Earth using a telescope. Photographs sent back from the US voyager spacecraft in the 1980 were able to identify further narrower rings in between the three main rings. The main rings are labeled A, B and C, with A the outermost ring. Recently more rings have been found. Saturn is the last planet that can be seen without using a telescope or binoculars and the planet was known in the ancient world before telescopes were invented. The rings, however, can only be seen using a telescope. The rings were first seen by Galileo in 1610 through a telescope. Saturn has at least 18 moon or satellites which orbit round the planet attracted to it by the planet's gravity. The largest of the moons, Titan, is the 2nd largest in the Solar system, after Jupiter's moon, Ganymede. Titan is larger than the Earth and is the only moon in the Solar system which is known to have an atmosphere. The atmosphere consists of nitrogen and methane. Titans were Jupiter's giant sons. Saturn itself is named, like all the planets, after a Roman God. Saturn was a rather mysterious God but it is believed that he was the God of sowing seed and of the harvest. Saturn is the second largest planet in the Solar System, after the giant Jupiter. Its mass is 95 times that of the Earth and it has a diameter of 75,098 miles, or 142,750 kilometers. Saturn is 886 million miles, or 1426 million kilometers, from the Sun. Saturn takes 29½ years to make one complete orbit of the Sun. The Earth takes one year.

Like Jupiter, however, Saturn spins much faster on its axis than the Earth. The Earth completes one rotation (turning) on its axis in 24 hours, turning away from the Sun and back again to give us night and day. Saturn, although so much bigger, completes a full rotation in just over 10 hours. This rapid spinning leads to hurricane-like storms far, far stronger than anything that is seen on the Earth. There is a constant whirlwind storm at Saturn's south pole which can be observed with the strongest telescopes.

The four largest outer planets, Jupiter, Saturn, Uranus and Neptune, are known as the "gas giants" since it is thought they are entirely made up of dense layers of gas. Saturn is a great ball of hydrogen and helium. Saturn's axis is tilted and as the planet orbits the Sun we get different views of the rings. Twice in every orbit only the edge of the outermost ring can be seen; even that can only be seen by using the strongest telescopes. Twice during the orbit we can see the fully opened rings. The rings all orbit Saturn at different speeds and have gaps between them. In 2010 the Cassini mission went between rings F and G and is now orbiting Saturn. The instruments on board this Cassini spacecraft are sending back valuable information which may help scientists to understand more about these mysterious and beautiful rings.

(Photos of Saturn)

Comets

A comet is an icy body that releases gas or dust. They are often compared to dirty snowballs, though recent research has led some scientists to call them snowy dirtballs. Comets contain dust, ice, carbon dioxide, ammonia, methane and more. Astronomers think comets are leftovers from the gas, dust, ice and rocks that initially formed the solar system about 4.6 billion years ago.

Some researchers think comets might have originally brought some of the water and organic molecules to Earth that now make up life here.

(Photo of a Comet)

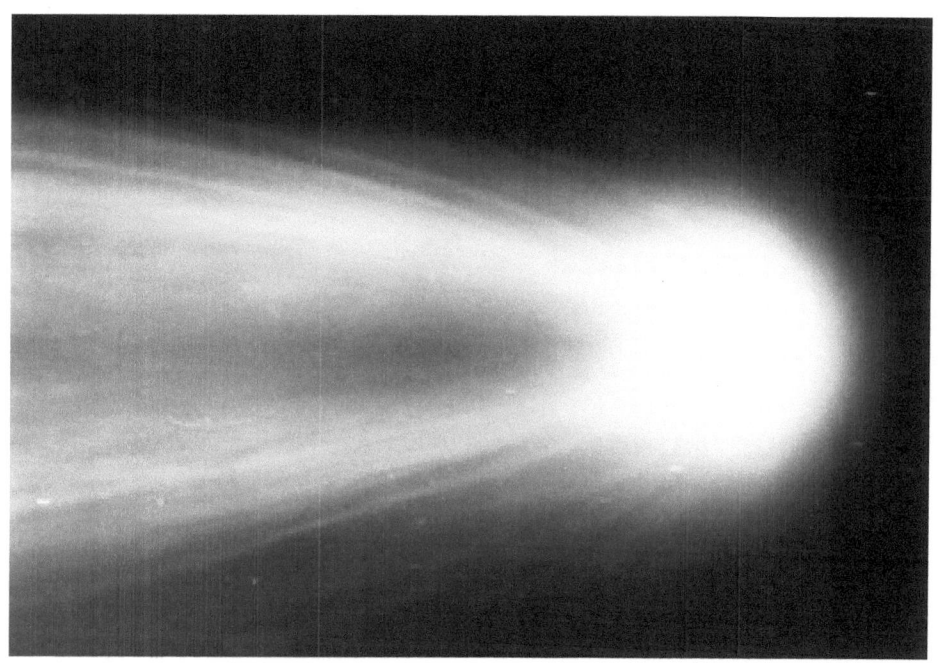

Uranus

Uranus cannot be seen from the Earth without a telescope. Uranus was first seen by William Herschel in 1781 during a survey of the sky using a telescope. Uranus has a total of 27 moons, most are named after characters in Shakespeare's Midsummer Night's Dream. The five major moons are called Titania, Oberon, Miranda, Ariel and Umbriel. Umbriel is not from Shakespeare but is the melancholy sprite in a poem by Alexander Pope. Most of the center of Uranus is a frozen mass of ammonia and methane, which gives it the blue-green color. The atmosphere also contains hydrogen and helium. Uranus orbits the Sun lying on its side and takes 84 years to complete one orbit. Because Uranus is lying on its side as it orbits the sun, for nearly a quarter of its orbit one pole of the planet is in complete darkness. Uranus takes 17.9 hours to turn once on its own axis, faster than the Earth, which takes 24 hours and gives us the change from day to night. Uranus was the ancient Greek God of the heavens whose sons were the Giants and Titans. Uranus is the smallest of the four giants, but is still several times larger than the Earth. It has a diameter of 29297 miles, or 47,150 kilometers, compared to the Earth's diameter of just under 8000 miles, or 12,760 kilometers.

Uranus is 1782 million miles, or 2869 million kilometers from the Sun. In 1977 Uranus passed in front of a star and astronomers observing the planet through giant telescopes saw nine rings around the planet. These are very faint and not easily seen, unlike the rings around Saturn. Photographs sent back by the Voyager spacecraft in the 1980's showed a further two rings round the planet. The Hubble space telescope found two more very faint rings, very distant from the planet, between 2003 and 2005, so we now know of a total of 13 rings. Scientists do not yet understand exactly what causes these rings or exactly what they consist of. From the time when Uranus was first observed scientists noticed that at certain points in Uranus' orbit the planet was being pulled further out into space. In the 19th century certain astronomers worked out that this must be because of the pull of gravity from another planet beyond Uranus. By making mathematical calculations based on the observations of Uranus, two astronomers, Adams and Le Verrier, identified where this other planet must be. The planet which was exerting a gravitational pull on Uranus was Neptune, 10.9 AU further out into space.

(Photo of Uranus)

Neptune

The discovery of the planet Neptune was one of the most exciting discoveries in astronomy. Neptune cannot be seen without a large telescope and was first seen in 1846 from the observatory in Berlin. But, the existence of Neptune had actually been discovered a year earlier, in 1845. Neptune is the 8th planet from the Sun. Uranus, the 7th planet, was first discovered by means of a telescope in 1781. Two astronomers J.C.Adams in England and Urbain Le Verrier in France, had been puzzled by the shape of Uranus' orbit. They worked out, using mathematics, that some large, more distant, body must be pulling Uranus towards it at certain points in the orbit. The Berlin observatory, following Le Verrier's calculations giving the possible position of this object, searched for Neptune and found the planet. They named it Neptune after the Roman God of the Sea. Neptune has a diameter of 29,297 miles, or 47,150 kilometers. Neptune is one of the four "gas giants". Like Jupiter, Saturn and Uranus, it is composed only of gas. Neptune is a great ball of hydrogen and helium. Like all the other planets in the Solar System, Neptune moves in an orbit round the Sun at the center of the system. It takes Neptune 165 of our Earth years to orbit the Sun. The Earth orbits the Sun in 365 days, one year. In 2011 Neptune completed the first orbit of the Sun since its discovery 165 years before in 1846. Like all the other planets, Neptune turns on its own axis as it is orbiting the sun but spins slightly faster on its axis, taking just over 19 hours to turn right round. Neptune is a staggering 2793 million miles (4495 million kilometers) from the Sun, and 2700 million miles from the Earth. In the same year that Neptune was first seen, 1846, its first moon was also spotted and named Triton. Triton orbits Neptune in the opposite direction of Neptune's own rotation on its axis. All the other major satellites or moons in the Solar System follow their planets round as they turn. We now know that Neptune has 13 moons in total. Since Neptune was the God of the Sea, all the moons are named after less important ancient Greek sea gods, like Triton, or sea nymphs, like Nereid.

(Photos of Neptune)

Pluto

Pluto was first seen by use of a telescope in 1930. Like Uranus and Neptune, Pluto can not be seen by the naked eye and its existence was not known to the ancient world. In 1930 the discovery of Pluto by Clyde Tombaugh in the Lowell Observatory was heralded as the discovery of the ninth planet. Even in 1930, however, many astronomers did not agree that a ninth planet had been discovered. In 2005 another possible planet, Eris, was found beyond Neptune, the eighth planet and since then 2 further small planet-like bodies have been discovered. These bodies are unlike the other planets in the Solar system. The first eight planets orbit the Sun in a path which they have cleared of other objects. Pluto, however, orbits the sun in a zone which is full of other objects that often pass between Pluto and the Sun. The first eight planets orbit the Sun while keeping the same distance from the Sun. Pluto orbits in an ellipse, an oval shape which means its distance from the Sun varies. In 2006 the International Astronomical Union agreed that these planets should be called dwarf planets. Pluto was the Roman God of the Underworld. Of the nine planets which most people think of as being in the Solar system, Pluto is the 2nd smallest, only just bigger than Mercury. It takes Pluto 248½ years to complete its orbit round the Sun. Like all the planets Pluto turns on its own axis as it orbits round the sun. Pluto takes about 6½ days to turn on its axis.

Since Pluto was the Roman God of the Underworld, the planet's main moon, Charon, is named after the ferryman who carries the dead souls across the River Styx into the Underworld. Pluto is known to have four moons. Pluto's distance from the Sun varies. Since the planet was only discovered in 1930 and it takes 249 years to orbit the Sun, a full orbit has not been observed. From calculations astronomers have worked out that Pluto's orbit round the sun is not regular. The orbit is tilted when compared to the orbits of the other eight planets Between 1979 and 1999 Pluto was closer to the Sun than the planet Neptune, moving inside Neptune's orbit.

(Photo of Pluto)

Makemake

Makemake is the largest of the Kuiper belt objects and the only one without satellites or moons. Although we know its diameter to be about 2/3 that of Pluto, its mass can merely be estimated in the absence of nearby objects. Makemake lacks its expected atmosphere. Astronomers thought Makemake would have developed an atmosphere similar to Pluto's, its chance passing in front of a bright star in 2011 revealed it mostly lacks a gas envelope. If present, Makemake's atmosphere would likely be methane and nitrogen-based.

Photo of Makemake

Eris

When Eris was first discovered in 2005, it was thought to be significantly larger than Pluto. Originally, it was submitted as the tenth planet in the solar system. Ultimately, however, Eris' discovery was a big reason astronomers demoted Pluto to dwarf planet status in 2006. That decision remains controversial to this day, making Eris' name fitting.

Eris is the Goddess of discord and strife. She stirs up jealousy and envy to cause fighting and anger among men. At the wedding of Peleus and Thetis, all the gods were invited with the exception of Eris, and enraged at her exclusion, she spitefully caused a quarrel among the goddesses that led the Trojan War. Observations helped scientists determine that Eris' diameter is 1,445 miles (2,326 kilometers), give or take 7 miles (12 km). That makes Eris' size even more precisely known than Pluto's. Pluto is thought to be between 1,429 and 1,491 miles — or 2,300 to 2,400 km — across.)

It also means that Pluto and Eris are, for all intents and purposes, the same size, researchers said. The researchers concluded that Eris is a spherical body. And, by studying the motion of Eris' moon Dysnomia, they peg the dwarf planet to be about 27 percent heavier than Pluto, which means it's considerably denser than Pluto as well. This density means that Eris is probably a large rocky body covered in a relatively thin mantle of ice.

Eris' companion

Eris is one of the few dwarf planets to boast a moon. Named Dysnomia, after Eris' daughter the demon goddess of lawlessness, the moon allowed astronomers to make more accurate measures of the planet than would have been otherwise possible, such as measurements of its density.

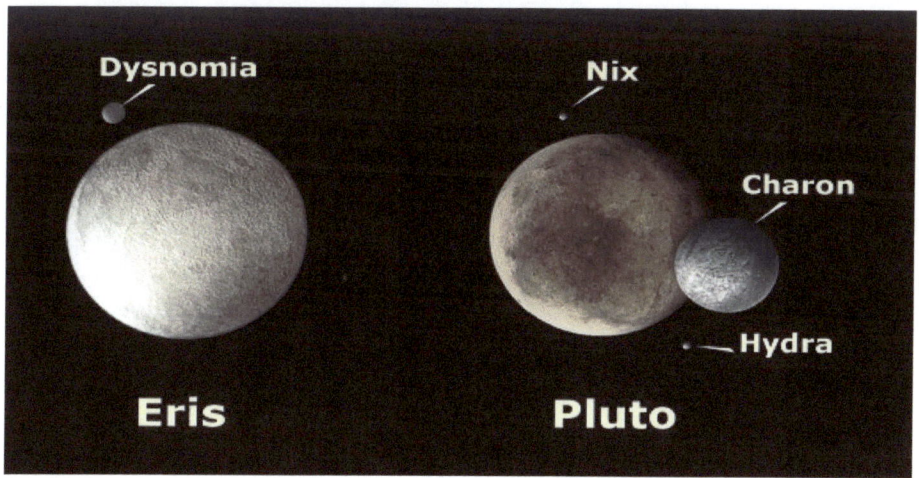

(Photo of Eris)

I hope you enjoyed reading and learning about our vast, beautiful and chaotic universe. There are so many things to learn, imagine one day we wake to find a distant race in a far away land, in another place.

Joni Morton - Lives in Cleveland Texas - Author of Children books, Myths and Legends, D I Y, Health and Beauty, and Recipe books. She is published in print and E-books through Amazon publishing.

~

~

Copyright © 2015 Joni Morton

All rights reserved. No part of the book may be reproduced in any form or by any electronic or mechanical means, including information storage and retrieval systems, without permission in writing from the author. Scanning uploading and electronic distributions of the book or the facilitation of such without the permission of the author is prohibited. Your support of the authors' rights is appreciated.